Effect of Foliar Spraying with Liquid Organic Fertilizer, some Micronutrients, and Gibberellins on Leaf Mineral Content, Fruit Set, Yield, and Fruit Quality of "Hollywood" Plum Trees

GRIN Publishing

Bibliographic information published by the German National Library:

The German National Library lists this publication in the National Bibliography; detailed bibliographic data are available on the Internet at http://dnb.dnb.de .

Imprint:

Copyright © 2010 GRIN Verlag GmbH
Print and binding: Books on Demand GmbH, Norderstedt Germany
ISBN: 978-3-656-94438-6

This book at GRIN:

http://www.grin.com/en/e-book/296179/effect-of-foliar-spraying-with-liquid-organic-fertilizer-some-micronutrients

GRIN - Your knowledge has value

Since its foundation in 1998, GRIN has specialized in publishing academic texts by students, college teachers and other academics as e-book and printed book. The website www.grin.com is an ideal platform for presenting term papers, final papers, scientific essays, dissertations and specialist books.

Visit us on the internet:

http://www.grin.com/

http://www.facebook.com/grincom

http://www.twitter.com/grin_com

AGRICULTURE AND BIOLOGY JOURNAL OF NORTH AMERICA
ISSN Print: 2151-7517, ISSN Online: 2151-7525
© 2010, ScienceHuβ, http://www.scihub.org/ABJNA

Effect of foliar spraying with liquid organic fertilizer, some micro-nutrients, and gibberellins on leaf mineral content, fruit set, yield, and fruit quality of "Hollywood" plum trees

Hassan, H.S.A.; S.M.A., Sarrwy; E.A.M., Mostafa

Pomology Department, National Research Center (NRC), Dokki, Giza, Egypt

ABSTRACT

This investigation was carried out during 2007 and 2008 growing seasons on "Hollywood" plum trees, grown in loamy clay soil condition at Sendyon village, Kalubia governorate, Egypt; Aiming to study the effect of foliar sprays with Aminofert (20% Amino acids, 12% organic acids, and 3.6% chelated micro-elements), gibberellins, and a mixture of chelated (Fe, Zn, and Mn) alone or in combination (GA_3 + Aminofert or GA_3 + a mixture of chelated "Fe, Zn, and Mn")on fruit set, yield, fruit quality, and leaf mineral content. Treatments increased significantly fruit set, yield as weight; or number of fruits/tree, as well as, fruit characteristics (Firmness, TSS, Flesh thickness, and Acidity) were improved under all treatments as compared to the control. Gibberellins or Aminofert alone or in combination (GA_3 + Aminofert or GA_3 + a mixture of chelated "Fe, Zn, and Mn") applied to foliage caused a pronounced increase in leaf N, and K content; while Leaf P content decreased in both experimental seasons. As for micro-nutrients concentration, the data indicated that the concentration of Fe, Zn, and Mn was icreased in all treatments comparing to the control. Application of Aminofert at 0.25% + GA_3 at 20 ppm was more effective compared to other treatments in the two seasons.

Keywords: Liquid organic fertilizer, Micro-nutrients, Gibberellins, plum, fruit set, yield, fruit quality, leaf mineral content

INTRODUCTION

Plums (*Prunus salicina*) are occupying an importance share in the total fruit production of Egypt. The total area of plum in Egypt reached about 2960 Feddans - according to the census of Ministry of Agriculture, Egypt (2005) - which produced 17148 tons with average yield of 5.79 tons / Fed.

Spraying of some nutrient elements (micro-nutrients), some growth regulators (GA_3), or liquid organic fertilizer (Aminofert) in order to increase fruit set, yield, and fruit quality of "Hollywood" plum. Mode of action for micro-elements was explained by Larue and Johnson (1989). Iron (Fe) complexes with proteins to form important enzymes in the plant and is associated with chloroplasts, where it has some roles in the synthesizing chlorophyll. Zinc (Zn) has been identified as component of almost 60 enzymes, therefore, it has a role in many plant functions, and it has a role as an enzyme in producing the growth hormone IAA. Manganese (Mn) participates in several important processes including photosynthesis , and metabolism of both nitrogen and carbohydrate..

On the other hand, foliar fertilizers as chelate should be easily absorbed by the plants, rapidly transported,

and shoud be easily release their ions to affect the plant Larue and Johnson (1989).

Amino acids have a chelating effect on micronutrients when applied together; the absorption and transportation of micronutrients inside the plant is easier, this effect is due to the chelating action, the effect of cell membrane permeability and low molecular weight Westwood (1993).

Foliar application seems to be effective in micro-nutrients deficiency symptoms El-Seginy *et. al,* (2003). Experiments showed an increase in fruit set and yield when gibberellic acid was applied to flower clusters Makarem & Mokhtar (1996), and El-Seginy & Khalil (2000). However, GA_3 treatment leads to reduce in fruit drop and improving most fruit characteristics Mansour (1979), Helail (1986), and Rezk (1988) on pear trees.

Also, the effect of GA_3 has at least three important actions, the first is intensify the ability of organ to be as a nutrient sink; secondly, increasing the synthesis of IAA in plant tissues; the third, it involves synthesis acceleration of hydrolytic enzymes in aleurone cells Addicott and Addicott (1982).

On the other hand, the effect of liquid organic fertilizer such as Aminofert was made for some purpose such as increasing fruit set, yield, and fruit quality; also, the harmful effect of using hormones can be avoided by using Aminofert. However, commercial application of liquid organic fertilizer at the orchard is still very limited El-Sayed (2005).

Accordingly, this study was aimed to evaluate the effect of spray application of a mixture of chelated (Fe, Zn, and Mn), gibbrillic acid, and Easterna Aminofert Super (Amino acids 20% + organic acids 12% + chelated micro-elements, Fe, Zn, and B 3.6%) on fruit set, yield, fruit quality, and leaf mineral content of "Hollywood" plum trees.

MATERIALS AND METHOD

This investigation was informed during 2007 and 2008 seasons on 15 years old trees of "Hollywood" plum cultivar grown in loamy clay soil at Sendyon village, Kalubia governorate, Egypt. Trees were cultivated at 5 X 5 m apart under basin irrigation system. Trees were healthy, similar in vigor and subjected to the same horticultural practices adapted in the region.

The selected trees were subjected to 6 treatments as follow:

1- Foliar sprays with tap water only (control).
2- Foliar sprays with mixture of chelated micro elements at 0.05%*.
3- Foliar sprays with GA$_3$ at 20ppm.
4- Foliar sprays with Easterna Aminofert at 0.25%.**
5- Foliar sprays with mixture of micro elements at 0.05% + GA$_3$ at 20ppm.
6- Foliar spray with Easterna Aminofert at 0.25% and GA$_3$ at 20ppm.

* A mixture of chelated composition:
Fe-EDTA (6%Fe), Zn-EDTA (14%Zn), and Mn-EDTA (12%Mn).

** Aminofert (liquid organic fertilizer) composition:
a- Amino acids: (Glycine, Glutamine, Systenine, Methionine, Aspartic acid, Gultamic acid, Valine, Lysine, Lecithin, Phynyl alanine) as 20 % of Aminofert composition.
b- Organic acids: (lactic, Citric, Aminobenzoic, Phosphoric, Acelic, Tartaric, Formic) as 12% of Aminofert composition.

c- Chelated microelements: (Fe 2%, Mn 0.5%, Zn 1%, and B 0.1%) as 3.6% of Aminofert composition.

Trees were sprayed with the above materials three times, at 70% full-bloom, after fruit set, and a month later. Foliar sprays were applied using a hand pressure sprayer. Triton-B emulsifier at a rate of 0.1% was used as a surfactant. Each tree received 5 liters of spraying solution; and two rows of trees were left surrounded each treatment as a guard border.

Treatments were replicated three times in a completely randomized block design. Each replicate consisted of three trees. The following parameters were determined in the two successive seasons:

Fruit Set: Two years old shoot was selected from each tree for recording data of total number of flowers at full-bloom in March and number of set fruits in April. These data were used in calculating the percentage of fruit set using the following equation:

Fruit set percentage =

$$\frac{Number \ of \ developing \ fruitlets}{Total \ number \ of \ flowers} \ x100$$

Yield per tree and fruit quality: Fruits were harvested at maturity stage (the first week of June) from each tree of various replicates and yield was recorded as a number of fruits/tree and weight in Kilograms. Samples of 10 randomly mature fruits from each experimental unit were used for measuring various fruit quality attributes. Characters measured were fruit weight, size, polar and equatorial diameters, flesh thickness, fruit firmness using a pressure tester, TSS content using a hand refractometer, and Titratable acidity percent as malic acid, A.O.A.C (1990).

Leaf mineral content: Leaf samples were collected for chemical analysis in early August of both seasons. Each sample consisted of 30 leaves / tree. Leaves were washed several times with tap water, rinsed with distilled water, and then dried at 70 č until a constant weight, ground and digested according Chapman and Pratt (1978). Nitrogen was estimated by semi-micro kieldahl method of Plummer (1971). Phosphorus was determined by the method outlined by Jackson (1973). Potassium, Fe, Mn, and Zn were determined using atomic absorption spectro photometer "Perkin Elmer 1100B" after samples digested according to Chapman and Pratt (1978).

The data were subjected to analysis of variance and the method of Duncan's (1955) was used to differentiate means.

RESULTS AND DISCUSSION

Data concerning the effect of treatments on fruit set % during the two experimental seasons are listed in

Table (1). The data cleared that, all GA₃ and Aminofert significantly increased fruit set percentage compared with the control at both seasons. Moreover, spraying GA₃ at 20ppm combined with liquid organic fertilizer (Aminofert) at 0.25% was more effective than the other trearments.

Table 1. Effect of spray with micro-nutrients, liquid organic fertilizer (Aminofert), and/or GA₃ on fruit set, and yield of "Hollywood" plum trees.

Treatments	Fruit set (%)		Yield (Kg/tree)		Number of fruits/ tree	
	2007	2008	2007	2008	2007	2008
Control	4.53 ^{c}	4.57 ^{c}	93 ^{d}	89 ^{d}	1974. ^{e}	1895 ^{e}
MC* at 0.05%	7.00 ^{b}	7.30 ^{b}	106 ^{c}	105 ^{c}	2159 ^{c}	2083 ^{d}
GA₃ at 20ppm	7.70 ^{ab}	7.90 ^{ab}	118 ^{b}	120 ^{b}	2428 ^{c}	2416 ^{bc}
Aminofert** at 0.25%	8.13 ^{ab}	8.30 ^{a}	130 ^{a}	131 ^{ab}	2519 ^{b}	2499 ^{b}
MC at 0.05% + GA₃ at 20ppm	7.97 ^{ab}	7.70 ^{ab}	128 ^{ab}	131 ^{ab}	2571 ^{a}	2407 ^{c}
Aminofert at 0.25% + GA₃ at 20ppm	8.47 ^{a}	8.53 ^{a}	138 ^{a}	142 ^{a}	2480 ^{c}	2544 ^{a}

*MC: a mixture of chelated Fe, Zn, and Mn.

**Aminofert: Liquid organic fertilizer (20% amino acid + 12% organic acid + 3.6% chelated micro-elements).

This result may be due to the use of plant growth regulators (GA₃) which could lead to an increase in fruit set of deciduous trees, Makarem and Mokhtar (1996). In addition, micronutrient elements are needed in relatively very small quantities for adequate plant growth and fruit production.

Generally, these results are in harmony with those reported by Makarem and Mokhtar (1996), Kabeel et al. (1998), and El-Seginy et al. (2003) when they worked on both pear and apple trees. Also, El-safty et al. (1998) and El-Sayed (2005) decided the same results on citrus and Abd-Ella and El-Sisi (2006) on fig trees.

Results as shown in Table (1) indicated that, spraying Aminofertat 0.25%, and GA₃ at 20ppm combined with Aminofert or chelated mixture exhibited favorable effect on increasing yield / tree (Kg) in the two experimental seasons. The lowest yield was recorded by control treatment followed by spraying chelated mixture (Fe, Zn, and Mn) at 0.05%, and GA₃ alone at 20ppm during both seasons.

The highest yield value was recorded by treatment Aminofert + GA₃ followed in decreasing order by

Aminofer, Mixture chelated + GA₃, mixture chelated alone, and control.

These results are owing to the use of GA₃ and micronutrients or Aminofert which led to an increase in fruit set and, GA₃ played a major role in enlarging fruit size.

In general, these results are in line with those obtained by Makarem and Mokhtar (1996), and El-Seginy and Khalil (2000). They reported that foliar spray of GA₃ increased fruit set, fruit weight, and as a result increased the yield. Also, El-Seginy et al. (2003), and Abd-Ella and El-Sisi (2006) found that foliar spray of GA₃ alone or combined with the chelated mixture (Zn, Mn, and Fe) increased the fruit set and total yield of Anna Apple trees and Sultani fig trees, respectively.

Results in Table (1) showed that, the chelated mixture (Fe, Zn, and Mn) at 0.05% + GA₃ at 20ppm exhibited favorable effect on increasing fruit numbers in the first season, while spraying Aminofert at 0.25% + GA₃ at 20ppm recorded the highest fruit number in the second season. Additionally chelated mixture (Fe, Zn, and Mn) or Aminofert treatments combined with GA₃ increased yield significantly than that of GA₃ or control treatments. Either the liquid organic fertilizer (Aminofert) alone or combined with GA₃ increased yield (as number of fruits /tree) as compared with the control. This supports the idea that Fe, Mn, organic and Amino acids sprayed on the leaves transported to leaves and fruits to improve nutritional statue and to avoid any yield depression, El-Sayed (2005).

Generally, these results are in agreement with the results obtained by Makarem and Mokhtar (1996), and El-Seginy and Khalil (2000); since they established that foliar spray of GA₃ increased the fruit set, fruit weight which lead to increase the yield. As well, El-Seginy et al. (2003), and Abd-Ella and ElSisi (2006) showed that foliar spray of GA₃ alone or in combination with the chelated mixture (Fe, Zn, and

Mn) increased fruit set and yield of Anna apple trees, and Sultani fig trees respectively.

Results obtained on fruit weight, size, and polar and equatorial dimension were affected significantly by different treatments. Spraying Aminofert at 0.25% alone or in combined with GA₃ at 20ppm significantly increased fruit weight, size, polar, and equatorial dimensions when compared with the control. Otherwise, spraying GA₃ at 20ppm with liquid organic fertilizer (Aminofert) at 0.25% was more effective than spraying GA₃ at 20ppm and chelated mixture of (Fe, Zn, and Mn) at 0.05%, or Aminofert alone at 0.25%. These results were accurate in both seasons (Table, 2). The results are conformity with those obtained by Eissa (2007).

Table 2. Effect of spraying with micro-nutrients, liquid organic fertilizer (Aminofert), and/or GA₃ on fruit weight, size, and dimensions of "Hollywood" plum trees.

Treatments	Fruit weight (gm)		Fruit size (cm³)		Polar diam. (cm)		Equatorial diam. (cm)	
	2007	2008	2007	2008	2007	2008	2007	2008
Control	46.87 ᵇ	47.10 ᵈ	45.23 ᵈ	48.10 ᵈ	4.45 ᶜ	4.31 ᶜ	3.90 ᶜ	4.20 ᵇ
MC* at 0.05%	48.47 ᵇ	49.67 ᶜᵈ	48.83 ᶜ	50.00 ᶜᵈ	4.83 ᵇᶜ	4.50 ᵇ	4.07 ᵇᶜ	4.26 ᵇ
GA₃ at 20ppm	49.10 ᵇ	50.40 ᶜ	49.50 ᵇᶜ	51.80 ᶜ	4.60 ᵇᶜ	4.55 ᵇ	4.14 ᵇ	4.30 ᵇ
Aminofert** at 0.25%	51.60 ᵃᵇ	52.30 ᵇᶜ	51.87 ᵇ	55.43 ᵇ	4.67 ᵃᵇ	4.61 ᵃᵇ	4.18 ᵃᵇ	4.48 ᵃ
MC at 0.05% + GA₃ at 20ppm	49.67 ᵇ	54.40 ᵃᵇ	50.53 ᵇᶜ	56.47 ᵃᵇ	4.61 ᵃᵇ	4.60 ᵃᵇ	4.26 ᵃᵇ	4.46 ᵃ
Aminofert at 0.25% + GA₃ at 20ppm	55.43 ᵃ	56.53 ᵃ	55.97 ᵃ	58.13 ᵃ	4.82 ᵃ	4.70 ᵃ	4.38 ᵃ	4.55 ᵃ

*MC: a mixture of chelated Fe, Zn, and Mn.

**Aminofert: Liquid organic fertilizer (20% amino acid + 12% organic acid + 3.6% chelated micro-elements).

.Table 3. Effect of spraying with micro-nutrients, liquid organic fertilizer (Aminofert), and/or GA₃ on fruit firmness, flesh thickness, TSS and titratable acidity of "Hollywood" plum fruits.

Treatments	Firmness (lb/inch²)		Flesh thickness (cm)		TSS (%)		Titratable acidity (%)	
	2007	2008	2007	2008	2007	2008	2007	2008
Control	5.90 ᵃ	6.37 ᵃ	1.44 ᵈ	1.62 ᵈ	11.20 ᶜ	10.80 ᶜ	0.60 ᵃ	0.58 ᵃ
MC* at 0.05%	4.44 ᵇᶜ	6.47 ᵃ	1.48 ᶜᵈ	1.70 ᶜᵈ	11.39 ᶜ	11.10 ᶜ	0.42 ᵇ	0.40 ᵃᵇ
GA₃ at 20ppm	5.37 ᶜ	5.57 ᵇ	1.54 ᵇᶜ	1.72 ᵇᶜ	11.98 ᵇᶜ	11.46 ᶜ	0.42 ᵇ	0.43 ᵃᵇ
Aminofert** at 0.25%	5.59 ᵇ	5.62 ᵇ	1.61 ᵇ	1.82 ᵃ	12.37 ᵃᵇ	12.37 ᵇ	0.48 ᵃᵇ	0.36 ᵃᵇ
MC at 0.05% + GA₃ at 20ppm	5.59 ᵇ	5.73 ᵇ	1.63 ᵇ	1.80 ᵃᵇ	12.13 ᵃᵇᶜ	12.73 ᵃᵇ	0.38 ᵇ	0.49 ᵃᵇ
Aminofert at 0.25% + GA₃ At 20ppm	5.64 ᵇ	6.12 ᵃ	1.74 ᵃ	1.86 ᵃ	13.00 ᵃ	13.43 ᵃ	0.35 ᵇ	0.33 ᵇ

*MC: a mixture of chelated Fe, Zn, and Mn.

**Aninofert: Liquid organic fertilizer (20% amino acid + 12% organic acid + 3.6% chelated micro-elements).

Fruit Quality: The data presented in Table (3) showed that, the experimental treatments at the two seasons generally increased fruit TSS, flesh thickness, and decreased fruit firmness, and juice acidity as compared with the control

These finding agreed with those found by El-Menshawi et al. (1997), El-safty et al. (1998), and El-Sayed (2005) on citrus. Also, El-Seginy and Khalil (2000), and El-Seginy et al. (2003) established the identical results on apple trees.

Leaf Nutrients Status: It's clear from the data in Table (4) that macro-nutrients leaf N% and K% content were significantly increased while leaf P% content was insignificant as results of spraying with micro-nutrients, liquid organic fertilizer (Aminofert) and/or GA₃ when compared with the control in both seasons.

Table 4. Effect of spraying with micro-nutrients, liquid organic fertilizer (Aminofert), and/or GA_3 on leaf mineral content of "Hollywood" plum trees.

Treatments	N %		P %		K %		Fe (ppm)		Zn (ppm)		Mn (ppm)	
	2007	2008	2007	2008	2007	2008	2007	2008	2007	2008	2007	2008
Control	2.19 d	2.07 c	0.26 b	0.26 c	1.00 c	1.02 d	80.00 d	82.00 d	41.00 c	44.00 c	37.00 d	39.00 d
MC* at 0.05%	2.41 c	2.41 b	0.28 ab	0.28 bc	1.08 bc	1.11 c	93.00 bc	95.00 c	47.00 b	49.00 b	43.00 c	45.00 c
GA₃ at 20ppm	2.48 c	2.49 b	0.33 ab	0.35 ab	1.15 ab	1.16 bc	91.00 bc	93.00 cd	46.00 b	48.00 b	42.00 c	42.00 c
Aminofert** at 0.25%	3.16 b	2.21 b	0.36 a	0.39 a	1.26 a	1.29 a	98.00 b	100.00 a	49.00 b	53.00 a	48.00 b	50.00 a
MC at 0.05% + GA₃ at 20ppm	2.44 c	2.46 b	0.30 ab	0.31 abc	1.15 ab	1.17 bc	95.00 bc	99.00 b	48.00 b	50.00 a	47.00 b	48.00 b
Aminofert at 0.25% + GA₃ At 20ppm	3.30 a	3.33 a	0.35 a	0.34 abc	1.17 ab	1.20 b	101.00 a	100.00 a	51.00 a	53.00 a	50.00 a	53.00 a

*MC: a mixture of chelated Fe, Zn, and Mn.

**Aminofert: Liquid organic fertilizer (20% amino acid + 12% organic acid + 3.6% chelated micro-elements).

Also, Aminofert at 0.25% alone or combined with GA₃ at 20ppm significantly increased N, and K % in the leaves than that of the control in the two seasons. As for micro-nutrients concentration, the data indicated that the concentration of Fe, Zn, and Mn, was increased in all spraying treatments. These results could be due to that GA₃ may intensify as organ ability to function as a nutrient sink Addicott and Addicott (1982).

The results approved to same extent with Sourour (1992), Kabeel et al. (1998), El-Seginy and Khalil (2000), El-Seginy et al. (2003), and Abd-Ella and El-Sisi (2006).

On the other hand, the status of nutrient in the plant resulted from spraying different solutions might be attributed to quick absorption via leaves and the limited loss of the nutrients when they were sprayed Marschner (1995).

These results agreed to same extent with El-Seginy and Khalil (2000), El-Shabaky et al. (2001), El-Seginy et al. (2003), and Abd-Ella and El-Sisi (2006) when they worked on deciduous trees, and El-Sayed (2005) who worked on citrus.

CONCLUSION: The obtained results in the present research strongly proved that foliar application of "Hollywood" plum trees with GA₃ alone at 20ppm, chelated mixture (Fe, Zn, and Mn) at 0.05% in combination with GA₃ at 20ppm, and Aminofert at 0.25% combined with GA₃ at 20ppm are recommended to increase fruit set, yield quantity, and fruit quality of trees grown under loamy soil condition.

Also, no significant differences were found between Aminofert and Aminofert combined with GA$_3$ treatments. Thus, the harmful effect of using growth regulators can be avoided by using such organic fertilizer or chelated mixture.

Otherwise, the combination of chelated form of micro-nutrients or Aminofert, with GA$_3$ appears to give the best responses.

REFERENCES

Abd-Ella Eman, E. K. and Wafaa A. A. Z. El-Sisi (2006): Effect of foliar application of gibberellic acid and micro-nutrients on leaf mineral content, fruit set, yield, and fruit quality of Sultani fig trees.. J. Agric. Res., Fac. Agric., Saba Basha, 11 (3):567-578.

Addicott, F. T. and A. B. Addicott (1982): Abscission UN. GA.. Press, Lts., London, England, pp. 30-135.

A. O. A. C. (1990): Official of analysis the association on official analytical chemists. 15th Ed, West Virginia, U.S.A, Washington D. C.

Chapman, H. D. and P. E. Pratt (1978): Methods of analysis for soils, plants, and waters. Univ. of Calif., Div. Agric. Sci., Priced Pub., 4034.

Duncan D. B. C. (1955): Multiple range and multiple (F) test Biometrics, 11: 1-24.

Eissa Fawzia, M. (2007): Enhancement of vegetative growth, flowering, yield, and fruit quality of "Hollywood" plum by using biostimulants and foliar fertilizers. J. Agric. Sci., Mansoura Univ., 32 (4):2759-2771.

El-Menshawi Elham, A.; H. M. Sinbel and Hoda A. Ismail (1997): Effect of foliar application of different zinc, manganese, and iron forms on yield and fruit quality of balady mandarin trees. J. Agric. Sci., Mansoura Univ., 22 (7):2333-2340.

El-safty M. A.; Elham A. El-Menshawi and M. A. Abd-Allah (1998): Effect of spray applications of GA and micro-nutrients on fruiting and leaf chemical composition of Washington navel orange trees. J. Agric. Res., Tanta Univ., 24 (2):208-214.

El-Sayed Ahalam A. (2005): Effect of foliar application of liquid organic fertilizer and/or GA$_3$ in fruiting and leaf mineral composition of Washington Navel orange trees. J. Agric. Res., Zagazig Univ., 32 (4):763-775.

El-Seginy A. M. and B. M. Khalil (2000): Effect of spraying some nutrients and gibberellic acid on leaf mineral content, fruit characters and yield of Le-Conte pear trees. J. Agric. Sci., Mansoura Univ., 25 (6):3529-3539.

El-Seginy A. M.; M. S. M. Naiema; W. M. Abd El-Messeih and G. I. Eliwa (2003): Effect of foliar spray of some micro-nutrients and gibberellins on leaf mineral content, fruit set, yield, and fruit quality of Anna apple trees. Alex. J. Agric. Res., 48 (3):137-143.

El-Shabaky M. A.; E. S. Abbas and H. A. El-Helw (2001): Effect of micro-nutrients spray on leaves mineral content, yield, quality, and storage ability of Ruby seedless grapes. J. Agric. Sci., Mansoura Univ., 26 (3):1721-1733.

Helail B. M. (1986): physiological response of young "Le-Conte" pear trees to some cultural and growth regulators treatments. PhD Thesis, Fac. of Agric., Moshtohor, Zagazig Univ., pp. 155

Jackson M. H. (1973): Soil chemical analysis. Prentice Hall. Inc., N. J.; Privatle Limited and New Delhi.

Kabeel H.; H. Mokhtar and M. M. Aly (1998): Effect of foliar application of different macro and micro nutrients on yield, fruit quality, and leaf mineral composition of Le-Conte pear. J. Agric. Sci., Mansoura Univ., 23 (7):3317-3325.

Laure J. H. and R. S. Hohnson (1989): Peaches, plums and nectarines growing and handling for fresh market. Copyright the Regent of the Univ. of Calif., Division of Agric. And Ntural Resources pub. 3331: 74-81.

Makarem M. M. and H. Mokhtar (1996): Effect of Biozyme and gibberellic acid on fruit set, yield, and fruit quality of Anna apple trees. J. Agric. Sci., Mansoura Univ., 21 (7):2661-2669.

Mansour A. S. (1979): Studies concerning the effect of gibberelic acid on pear. M. Sci. Thesis, Fac. of Agric., Ain Shams Univ., Cairo, Egypt, pp. 110.

Marschner H. (1995): Mineral nutrition of higher plants. Second Edition; Academic press. London.

Ministry of Agric., A.R.E (2005): Acreage and total production of Agric. Crops in A.R.E. Bull. Agric. Econ. and Statistics (In Arabic).

Plummer D. T. (1971): an introduction to practical biochem. Mc-Graw Hill Book Company; U. K. limited.

Rezk M.H. (1988): Effect of some growth regulators on pear trees. M. Sci. Thesis, Fac. of Agric., Mansoura Univ., Egypt, pp. 96.

Sourour M. M. (1992): Response of Anna apple trees to different methods and forms of iron applications. Alex. J. Agric. Res., 37 (2):191-204.

Westwood M. N. (1993): Temperate – zone pomology physiology and culture. Third Edition. Himber press, Portland, Oregon, p. 523.